Animal Units

Jo Ellen Moore • Joy Evans

"Teaching Drawing For The Artistic Klutz And Others"

by
Evans and Moore

"I CAN'T DRAW!" This was the response we got from many of our fellow teachers when we asked them why they didn't emphasize art activities in their classrooms. One of our colleagues commented in a faculty meeting, "It was after I tried teaching my class how to draw a doggie that I seriously questioned how I could have earned a liberal ARTS degree!" Unfortunately, art has become a rainy day activity for many of us when it could be a vital part of our elementary curriculum.

Many teachers also feel discouraged because their students lack artistic confidence. "I can't draw... they can't draw. Maybe we can just stick to crayons and string pull paintings again this year." NONSENSE! You can draw... they will learn to draw. There's no big trick to it.

Using the techniques in this book, we have found that even first grade students become confident and comfortable with a wide variety of art projects. Fear of failure (for you and the student) can be replaced with the excitement and joy of a successfully completed project. This, in turn, generates a willingness to experiment.

There are many positive side effects to a super-charged art program, to wit:

- **greater small muscle control**
- **improved ability to follow auditory and visual directions**
- **expanded powers of observation; and**
- **a better self-image.**

In some cases where children had serious academic and social problems, success in drawing has been a first step toward a solution.

A WORD OF CAUTION: Throw caution to the wind and ENJOY!

1. You may draw these lessons <u>directly</u> on the glass of your overhead projector with overhead pens and then wipe off with a damp rag. (Do <u>not</u> use pens with permanent ink!)

2. Children must have a clear view of the screen and be encouraged to listen attentively.

3. Be conscious of the vocabulary that can be developed in each lesson.

4. Turn off the projector to recall children's attention.

5. Pick out of each lesson what works for you. Each unit has a drawing lesson, poetry, language arts ideas, suggested readings, science facts, bulletin boards, and 3-D art lessons.

©1979 by EVAN-MOOR CORP.

CONTENTS

	PAGES
Caterpillar Capers	6
Snail Trails	14
Turtle Topics	22
Frog Funnies	30
Crocodile Wiles	38
Mouse Mischief	46
Walrus Whimsy	54
Bear Fare	62
Elephant Illustrations	70
Giraffe Acts	78

Caterpillar Capers

Produce the drawings in steps on an overhead projector or chalkboard as the child follows on his paper.

Keep talking as you draw.

What shape am I beginning with?
I begin my line on the left and go across the page.
Draw a small circle inside the larger one.
This line is a diagonal.

Develop descriptive vocabulary related to each animal that has been highlighted by reading to the class the suggested poetry and literature selections.

Encourage children to draw LARGE enough to fill the paper.

Drawing Steps

Draw circles from left to right.
Segments move up and down.
Circles are the same size.

Facial Expression:
- button nose
- sleepy eye with drooping lid
- grinning mouth in profile

Each segment needs a foot.
Is he barefoot or booted?

Add a floppy hat and fragrant flower.
The trail begins on the left and swells up and down to meet his feet.

Is your caterpillar:

    diagonal   striped   zig-zagged
    crossed   wavy   polka dotted

So now you've drawn a caterpillar.
Now what? How about...

    Outlining with black   Adding the background   Coloring the fellow brightly

Follow up the next day by asking children to draw a caterpillar looking for his lunch.

Language Art Starts

Today...
A hungry caterpillar
Feasts on leaves
　　up high
Tomorrow...
A moth or butterfly
Silently flies by.

　　　J.E. Moore

WHAT makes me fuzzy?

WHEN is it time to eat?

WHY do I eat leaves?

Caterpillars can be a lively topic for oral or written language experiences. Try developing unusual WHAT, WHEN and WHY questions.

Literature

Sphinx: The Story of a Caterpillar
　　by Rubert McClung
　　Pictures by Carol Lerner
Morrow

The life cycle of a caterpillar is told in story form.

Look...A Butterfly
　　by David Cutts
　　Pictures by Eulala Conner
Troll

A simple introduction to butterflies for younger students.

*It's Easy to Have
a Caterpillar Visit You*
　　by Caroline O'Hagan
　　Pictures by Judith Allan
Lothrop

Children learn how to find caterpillars and how to take care of them.

Terry and the Caterpillar
　　by Millicent E. Selsam
　　Pictures by Arnold Lobel
Harper and Row

An easy reading story full of facts on raising caterpillars at home.

The Hungry Caterpillar
　　by Eric Carle
Collins

A hungry little caterpillar eats its way through a multitude of plant forms.

Caterpillars
　　by Dorothy Sterling
　　Pictures by Winifred Lubell
Doubleday and Co., Inc.

Gives facts on growth of caterpillars and where to find and how to grow them. Also, information is given on silk worms and garden pests.

Caterpillar Facts

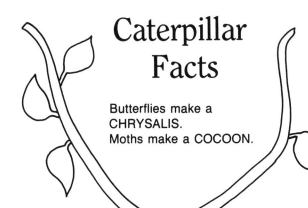

Butterflies make a CHRYSALIS.
Moths make a COCOON.

Baby caterpillar's first meal may be his own eggshell.

It's tiny caterpillars inside Mexican jumping beans that make them jump.

Polyphemus Moth caterpillar increases its weight more than 4,000 times.

Their skin doesn't grow. They shed old skin when outgrown.

Unbored Bulletin Boards

Paint a vine directly on butcher paper covering the bulletin board space. Pin on the 3-D caterpillars and have the children write a "caterpillar quote" on paper cut in a bubble shape.

Terrarium Fun
Let sharp-eyed children catch caterpillars for the school terrarium. Cover with screening and provide leaves from pet's original tree.

3-D art stuff	Egg Carton Caterpillars	Materials: • 1 egg carton bottom • 2 pieces of pipe cleaner • felt marking pens

1 Cut the bottom part of the egg carton in half the long way.

2 Decorate five of the sections with felt pens. The sixth segment gets the face.

3 Poke the pipe cleaners into the head section. What personality!

©1979 by EVAN-MOOR CORP.

Draw a caterpillar looking for its lunch.

Using words from the Word Bank, write about your picture.

Word Bank

small	dots	crawl	hungry
long	segments	climb	cocoon
fuzzy	smooth	grow	butterfly
color	munch	change	moth
stripes	nibble	hatch	caterpillar

In each circle put a word which tells about caterpillar!

This is _____'s caterpillar.

What does it look like?
What can caterpillar do?
How does it move?

Dear Parents,

We have just completed marvelous caterpillars made from egg cartons. Hidden in the throw-aways from your kitchen are other fanciful animals. Encourage your child to create a new creature from these scraps. Then bring the new creation to school to share with all of us.

Thank you.

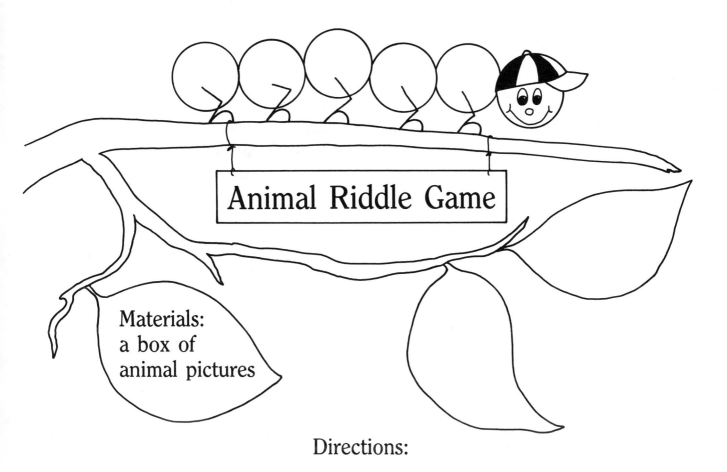

Animal Riddle Game

Materials: a box of animal pictures

Directions:

Child selects a picture from the box and describes the animal chosen.

- Appearance
- Way It Moves
- Something Unique It Does

Classmates attempt to guess the animal. If no one is correct after 5 guesses, the child shows his picture and selects someone to take his place.

If someone guesses correctly, he becomes "It."

Snail Trails

Direct the drawing lesson on the overhead projector by drawing directly on the glass with an overhead projector pen. You may also use the chalkboard effectively with these lessons.

The class follows each step on their large drawing paper. It helps to encourage them to sketch lightly with their pencils.

Every shape you draw can be described. Lines can be curvy, wavy, long, longer, etc. This can be a language lesson as well as art. It is designed to equally develop language skills and visual awareness.

Drawing Steps

Draw the large circle in the center of the paper. Begin the spiral on the bottom of the circle and roll inward.

Now sketch in the head and tail. Remember to tell the children that their drawing doesn't have to be identical to yours.

All we need here is personality. Our snail has his basic parts, so now we can diversify. Facial features:

Begin on the left-hand side and add the dotted line for the snail's trail. It can go up, over, around or anywhere.

Let's think about where this snail is spending the afternoon. Now add your ideas to your drawing so it reflects you.

 Perhaps he's crawling over a log?

Eyeing a juicy petunia?

 Basking in the sun?

POST TEST

The child is to follow up the next day by drawing a snail upside down on a leaf.

Language Art Starts

People call me pokey
Because I move so slow
But if you had just
   one foot
How fast would YOU go?

        J.E. Moore

A snail makes a terrific classroom pet. We're easy to care for, fascinating to watch and fun to race!

Snail Tales

Cut writing paper into a round snail shape. Paste it on a 12" x 18" piece of white construction paper. Write a "snail tale" on the shell-shaped paper. Use crayons to draw snail's head, tail and background.

Literature

Snail
   by Jens Olesen
   Pictures by Bo Jarner
   Silver Burdett

The text and photographs examine the characteristics and behavior of the snail and follow its life cycle.

Snail, Where Are You?
   by Tomi Ungerer
   Harper and Row

A good book for Kindergarten or first grade. There is no story, just delightful pictures using a snail shell as part of some item on each page.

The Biggest House
   by Leo Lionni
   Pantheon Books

The tale of a little snail determined to have the most beautiful shell in the world.

The Snail's Spell
   by Joanne Ryder
   Frederick Warne

Many facts about snails are taught in this fantasy story about a boy who shrinks down to the size of a snail and wanders through the garden.

Snails
   by Dorothy Childs Hogner
   Pictures by Nils Hogner
   Thomas Y. Crowell

This book covers parts of a snail's body, reproduction, behavior, enemies, and how to keep them as pets.

Unbored Bulletin Boards

Snail Facts

- Some snails are as small as a thimble; some have 5-pound shells.
- Snails have been around for millions of years.
- A gland on the foot secretes a slime which facilitates movement.
- Snails cannot hear or make sounds.
- They are found everywhere: mountains, valleys, and oceans.
- They detect changes in light but not form.

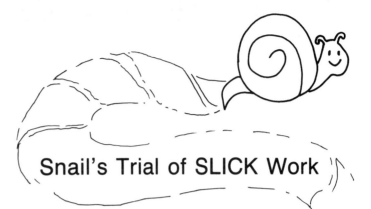

Snail's Trial of SLICK Work

Snail's shell is cut from big brown bag.
The "trail" is twisted Saran warp.
The caption is printed on butcher paper.
Pin up children's work along trail.

Some people eat snails! Would you?

3-D art stuff — Paper Snails

Materials:
- construction paper strips 2" (5 cm) wide brown (18", 12", 8" long) (45.7, 30.5, 20.3 cm) yellow (15" long) (38 cm)
- pipe cleaners (2 per snail—one inch)
- paste, crayons and scissors
- 9" x 12" green paper (22.9 x 30.5 cm)

Directions:
This lesson may be varied to fit the abilities of your class. For young children, the papers may be pre-cut or done on a ditto master and run on white art paper.

Older children could use rulers to measure strips.

Some children would enjoy decorating strips with zig-zags, polka-dots and stripes before pasting circles.

1. Form the brown strips into circles
   8" strip 12" strip 18" strip
   Paste
2. Paste the brown circles inside one another.
   Paste at base
3. Cut yellow strip for the snail body.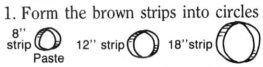
4. Curl the head and the tail on a pencil.
5. Paste the shell in the center of the body.
6. Paste on the pipe cleaner antennae.
7. Paste entire snail on the green 9" x 12" paper. Draw some luscious leaves!

©1979 by EVAN-MOOR CORP.   Animal Units

Draw a snail upside down on a leaf.

Using words from the Word Bank, write about your picture.

Word Bank

large	soft	climb	shell
round	pokey	move	slime
curve	hard	eyestalks	trail
design	explore	creep	foot
colorful	gobble	crawl	leaf

This animal is very busy.
What is he doing?
Please draw him so
   that I can see
   what is happening.
Thank you!

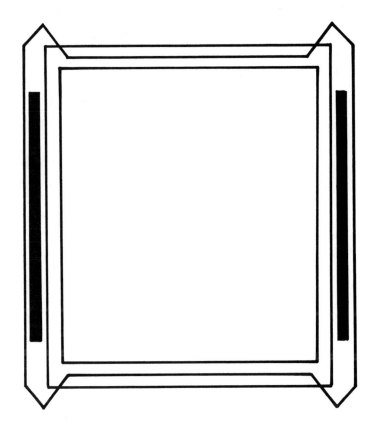

Describe this animal. Use three words.

How does this animal move?

Where is it going?

When is this happening?

Put all these answers into one good sentence!

Dear Parents,

A snail makes an interesting temporary pet for your child. All that is needed is a jar, a lid with air holes, a smaller lid with water and a snail from the yard.

Encourage your child to watch carefully to find out what snail eats and how it moves. A magnifying glass is great for a closer look. Watch with your child. You may find it fascinating too.

Thank you.

P.S. Other backyard creatures can be great pets too. How about earthworms or insects?

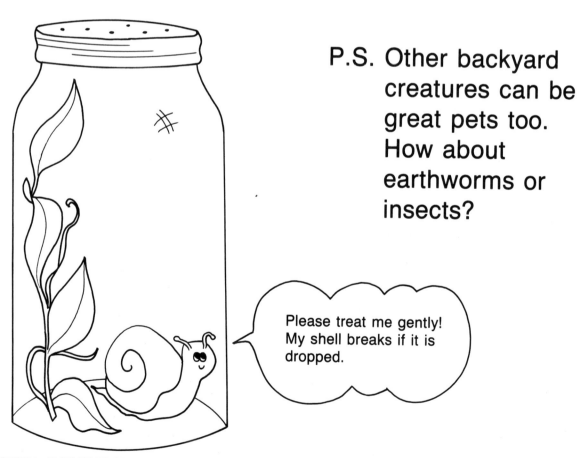

Please treat me gently! My shell breaks if it is dropped.

Be a HAWK.....

Move Along

This game should be played indoors or on a grassy area where children will be free to get on their hands, knees and stomachs.

Teacher names an animal.
The class tries to show how that animal moves.
Include animals that crawl, slither, walk, run, fly, hop, swim, jump, etc.

See my trunk?

My antennae can twitch!

Turtle Topics

By now the students in your class know how to follow and sketch as you lead on the overhead or chalkboard.

Remember to keep talking as you draw. You are developing creative vocabulary and exciting science facts as you teach the children how to produce the animal's basic shape on paper.

The creative instincts in your children will take these basics and strike out in many different directions. You'll find this is only the beginning!

This lesson introduces what it means to draw a figure in profile.

Continue to encourage drawings large enough to fill the paper.

©1979 by EVAN-MOOR CORP. Animal Units

Drawing Steps

Begin with a straight line below the center of the paper and off-center to the left. Draw a large hill above the line.

Turtle's head now appears on the right and the somewhat bent tail on the left. Proportions here are flexible; don't fret!

We are drawing the profile of this turtle; so only two paddle-like feet appear. His toes add interest and design.

Profile facial expressions may vary. Pick one!

- eyes
- smile / frown
- eyebrows
- cheeks

Turtle's shell can have any pattern.

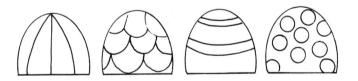

Now the children should be encouraged to add some finishing touches.
- a jaunty hat
- a bouquet in his mouth
- a bird nesting on his shell

Follow up soon with a "draw your own turtle" lesson just to practice some different shell designs.

DESIGN IT

Got any bright ideas?

POST TEST

Unbored Bulletin Boards

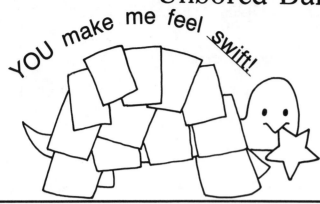

This makes for a happy place to rest your eyes during the day. Children's daily work is pinned on the turtle's shell. It builds self-esteem in the students and is a great interest getter. Another option is to place "Turtle Facts" in each section of the shell so that it can be instructive as well as attractive.

Turtle Facts

- Alligator Snapper Turtle has a long pink ridge on its tongue to attract fish. He stays on the bottom of the pond with his mouth open until a fish appears. Then SNAP!

- Turtles have NO TEETH!

- Turtles bury their eggs in holes on land.

- Box Turtle has a shell so strong it can support 200 times its own weight.

- Some turtles live over 100 years.

- Some turtle eggs are edible.

- Turtles may weigh 500 pounds.

Literature

Turtle Pond
  Berniece Freschet
  Pictures by Donald Carrick
  by Scribners

This story centers on a turtle and her eggs as it describes life in a pond.

Let's Get Turtles
  by Milicent E. Selsam
  Pictures by Arnold Lobel
  Harper and Row

An easy-to-read text written in story form helps children learn many facts about turtles and how to take care of them.

The Tortoise and the Hare
  An Aesop Fable
  by Paul Galdone
  McGraw

A charming re-telling of the race between the tortoise and the hare.

Something New for Taco
  by Jane Castellanos
  Pictures by Bernard Garbutt
  Golden Gate Junior

Adventures of a desert tortoise when it rains.

What is a Turtle?
  by Gene Darby
  Pictures by Lucy and John Hawkinson
  Benefic Press

Many colorful pictures of the turtle life cycle and of its eating habits.

Language Art Starts

Turtles

Little Turtle's home
Is never far away
He carries it along
As he goes about his day.

As little turtle grows
His house grows too
It fits him as well
As a shoe fits you.

— J.E. Moore

Challenge Students to Answer Questions

- How are snails and turtles alike? How are they different?
- See how many animals with a shell you can name in one minute. Can you draw them?
- Use a dictionary or encyclopedia to discover the difference between turtles and tortoises.

3-D art stuff — Turtle on a String

Materials:
- 2 paper plates (Size is optional.)
- 6 egg carton cups
- 25 inches of string or yarn (61.5 cm)
- construction paper scraps
- scissors, glue, crayons, hole punch

Directions:
This lesson uses materials readily available in classrooms. The result is an endearing turtle that can be pulled along the ground just like a real pet. The lesson can be <u>adapted</u> to all levels.

1. Cut legs and tail from scrap paper and glue on edge of plate.

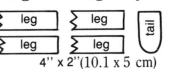
4" x 2" (10.1 x 5 cm)

2. Paste 2nd plate over the top.

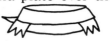

3. The head is a 2" x 6" (5 x 15.2 cm) paper strip. Fold in half, draw eyes and mouth with crayons. Glue on plate.

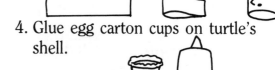

4. Glue egg carton cups on turtle's shell.

5. NOW: paint him with tempera if you wish or just punch a hole in the plate under his chin and add a pull string. He will follow anywhere!

©1979 by EVAN-MOOR CORP.

Draw a turtle with a super shell design.

Using words from the Word Bank, write about your picture.

Word Bank

bumpy	happy	turtle	shell
smooth	surprised	tortoise	head
pattern	crawled	eggs	feet
pokey	lay	bury	tail

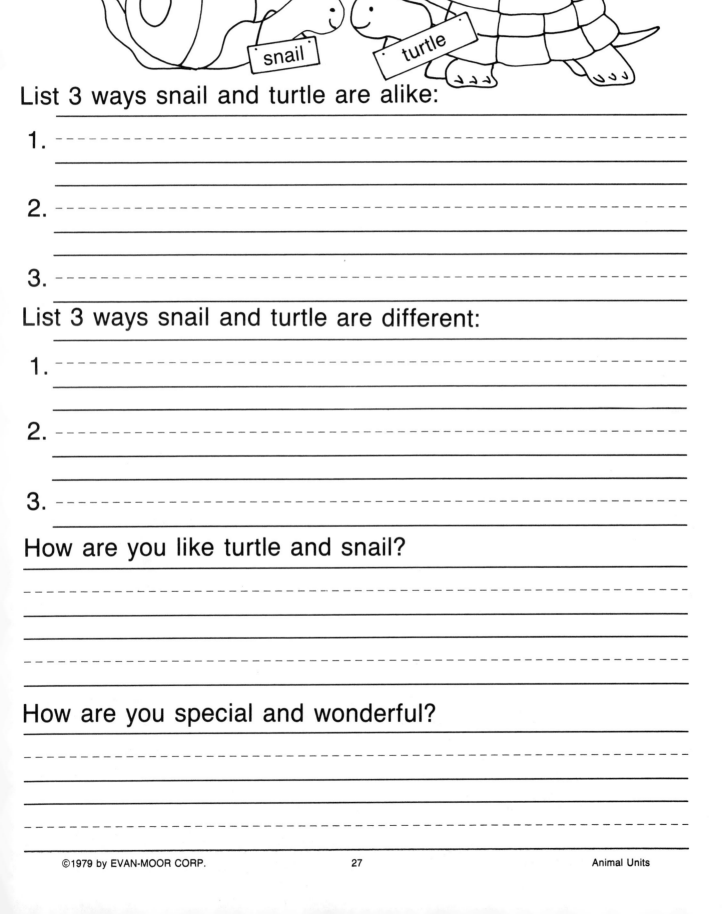

List 3 ways snail and turtle are alike:

1. _____

2. _____

3. _____

List 3 ways snail and turtle are different:

1. _____

2. _____

3. _____

How are you like turtle and snail?

_____

_____

_____

How are you special and wonderful?

_____

_____

_____

How many animals with a shell can you name?

Write the names below.

1. _____
2. _____
3. _____
4. _____
5. _____
6. _____
7. _____
8. _____
9. _____
10. _____

See if you can find pictures of any of these shell animals.

Look in newspapers and magazines.

Paste the pictures on the back of this paper.

Tortoise and Hare

Play outdoors, in a gym or other area where you have space.

Have the children form a circle all facing clockwise.

When the teacher or a child chosen to be "It" calls TORTOISE, everyone moves as slowly as possible. When "It" calls HARE, everyone races quickly around the circle.

Variation: Make the game more difficult by having children move slowly when any slow animal is named (snail, worm, etc.) and quickly when fast animals are announced (horse, cheetah, etc.).

This would be a good time to read **The Tortoise and the Hare** from Aesop's Fables.

Frog Funnies

Frog offers a good lesson in change of facial expression. The movement of eyes and mouth are easy to draw and they create a new feeling each time. Movement can also be simulated by minor changes in feet positions.

As you work on the overhead projector or chalkboard you are developing an easy-to-draw character to use in creative writing stories or perhaps even a "comic strip" will evolve around his escapades.

This lesson also develops simple perspective concepts.
    1. Objects in the distance appear smaller.
    2. Distant objects will be closer to the top of the paper.

Comparative adjectives like large, larger, largest are useful here.

Drawing Steps

Got any bright ideas?

POST TEST

Three horizontal lines have graduated lengths. The longest one is closer to the bottom of the page because it is nearest.

Now we have three hills drawn over the lines. They appear to recede because of size differences and placement on the page.

The frogs' energetic jumping legs are formed by smaller hills drawn at the base of the larger one.

The bulbous eyes perch above a broad smile. Facial expressions vary easily and inventively!

The frogs' smaller front legs help to stabilize him on hazardous landings. They have four toes. His webbed back feet are ready to jump.

What's your frog doing?
- Resting upon a lily pad?
- Hopping around in search of BUGS?
- Snuggled in for a long winter's nap?

Draw a progression of frogs with their legs in a new position each time to create movement.

Frogs
   leaping away with
     smooth and shiny skin
Toads
   hopping along
     are frog's "bumpy" kin.

          J.E. Moore

Frog Facts:

- Frogs are amphibians. (They live part of the time in water and part of the time on land.)
- They are cold-blooded.
- Frogs hibernate in winter.
- A frog's tongue is attached to the front of its mouth and is sticky.
- They help man by eating insects.
- Baby frogs are called tadpoles or polliwogs.
- Frogs have **no** necks.
- Some frogs grow to 10 inches.
- Frogs have smooth, moist skin.
- Toads have dry, rough, bumpy skin and larger eyes.

Literature

Spring Peepers
   by Judy Hawes
   Pictures by Graham Booth
   Harper

A simply written description of the life of a tree frog.

The Frog in the Well
   by Alvin Tresselt
   Pictures by Roger Duvoisin
   Lothrop, Lee & Shepard

The story of a frog forced to leave his home at the bottom of a well when it dries up.

Frog Went A-Courtin'
   by John Langstaff
   Pictures by Feodor Rojankowsky
   Harcourt Brace

An old-fashioned version of the familiar song of Frog and Miss Mousie.

Frogs and Toads
   by Jane Dallinger
   and Sylvia Johnson
   Lerner

This book explains the life cycle of toads and frogs.

A Frog's Body
   by Joanna Cole
   Morrow

An excellent introduction to the life cycle and anatomy of a bullfrog.

Frog and Toad Are Friends
   by Arnold Lobel
   Harper & Row

Easy-to-read misadventures of those good friends Frog and Toad.

Let's Find Out About Frogs
   by Corinne J. Naden
   Pictures by Jerry Lang
   Franklin Watts, Inc.

A fascinating and clear presentation of facts about frogs.

Language Art Starts

Who	Problem	Solution

This is a basic lesson to help children THINK through a story plot. Give each child a ditto showing three boxes labeled: Who, Problem and Solution. Using a frog as the Who, have the children develop an exciting Problem and a corresponding Solution. They illustrate the strip and use bubbles to provide dialogue. Older children may then want to go on to a more detailed comic strip using more boxes.

Unbored Bulletin Boards

A Compound Lunch

Hungry frog **won't** be overlooked on your bulletin board. His body is simply formed from green paper following the form in the drawing lesson. Make him as big as your bulletin board allows.
His tongue is a strip of crepe paper. The insects are formed from circles. The title is cut out of black paper.

3-D art stuff — Leaping FROG

Materials:
- 1 sheet 9" x 4" (22.9 x 10.2 cm) paper (any color)
- 2 white ½" (1.3 cm) circles
- paste and crayons

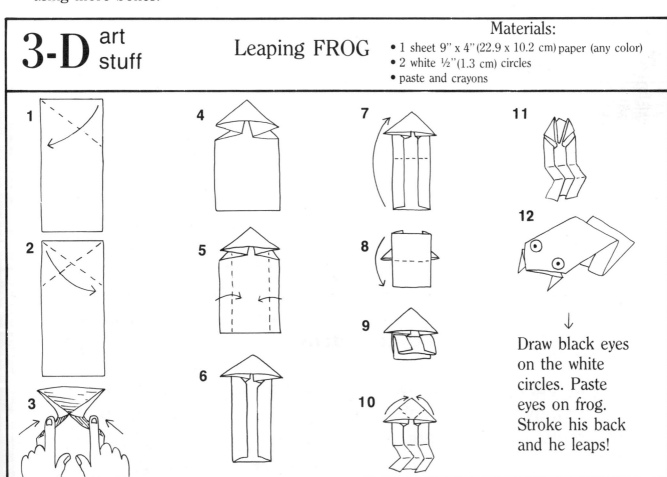

Draw black eyes on the white circles. Paste eyes on frog. Stroke his back and he leaps!

©1979 by EVAN-MOOR CORP.   Animal Units

Draw three frogs hopping along.

Using words from the Word Bank, write about your picture.

Word Bank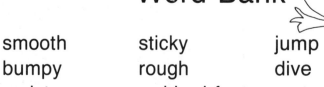

smooth	sticky	jump	tadpole
bumpy	rough	dive	polliwog
moist	webbed feet	rest	toad
wet	cold	tongue	frog

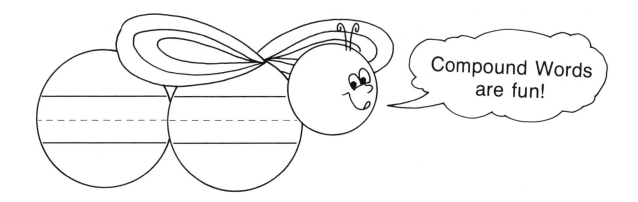

Compound Words are fun!

Choose a word from each Word Bank.
Write them in the circles to make compound words.

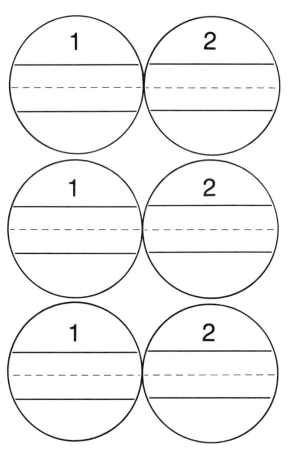
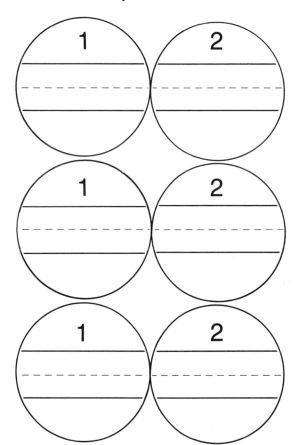

Word Bank 1		Word Bank 2	
cup	mush	bow	cake
rain	pine	room	plane
air	sail	apple	boat

Frogs
   leaping away with
   smooth and shiny skin
Toads
   hopping along
   are frog's "bumpy" kin.

Answer these questions about frog and toad:

1. Which word tells how frog moves? _____

2. Which word tells how toad moves? _____

3. Which word tells about toad's skin? _____

4. Two words tell about frog's skin.
   Write them.
   _____   _____

5. What does kin mean?
   _____

Feed the Frog King

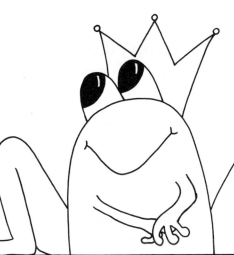

Choose one child to be the scorekeeper. This child becomes the Frog King (or Queen).

Each team of 2-4 players has an envelope of insect parts containing words. Each team tries to match up an insect front and back to create a compound word. The "finished insect" is delivered to the Frog King and the team receives a point for each correct answer. (Use insect pattern on compound word exercise page.)

Play continues until all words are used, or you can set a time limit.

Sample words:
- cupcake
- ballgame
- policeman
- tadpole
- butterfly
- jellybean
- tugboat
- football
- seaweed
- popcorn
- cowboy
- maybe

Crocodile Wiles

Before you begin this lesson, be sure you know what makes a crocodile different from an alligator. There is always a child in class who knows the facts. The crocodile fact section will help you defend yourself.

This directed drawing lesson again emphasizes:
- Sketching lightly with pencil.
- Drawing large figures that fill the page.
- The development of descriptive vocabulary.

The crocodile's zig-zagged tail and snout, the swamp grass in the background, and cross hatching relief on his back all provide good design elements for adding interest and variety to the completed drawing.

Drawing Steps

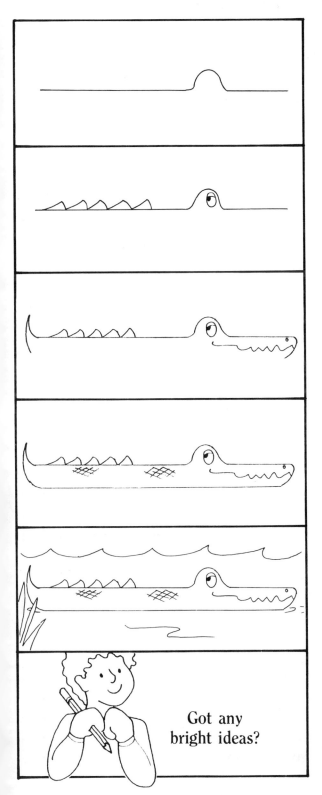

Got any bright ideas?

POST TEST

Begin on the left side below the midline of the paper. The hill rises up just after passing center.

The tail is zig-zagged. There is a sleepy eye watching you.

The smile curls around below the eye. A powerful tail curves up.

Draw the lower line parallel to the top one. Add cross hatching on the crocodile's back.

Draw water lines around the crocodile.

Is your crocodile looking for a tasty fish snack?
Does your crocodile have a bird on his head?

Follow up by having children independently draw crocodile basking in the sun.

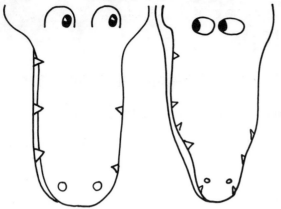

Language Art Starts

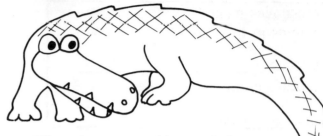

Alligator has a broad, broad nose
Crocodile's is thin
But both of them are dangerous
When they begin to grin.

       J.E. Moore

What does crocodile need that tremendous tail for anyhow? After reading factual books to the class on crocodile habits, encourage them to name as many possible uses as they can. Ideas may vary from fact to fiction and may even lead to full length escapades rather than the one line provided.

Literature

Reptiles Do the Strangest Things
    by Leonora & Arthur Hornblow
    Pictures by Michael K. Firth
    Random House

An easy-to-read science book. It covers many reptiles, but has an excellent chapter on crocodiles and alligators.

Alligator
    by Evelyn Shaw
    Pictures by Frances Zweifel
    Harper

A simple description of the alligator's life cycle.

Who Needs Alligators?
    by Patricia Lauber
    Garrard

A description of alligators, their habits, homes, reproductive cycle, and the animals that prey on them.

Alligators and Crocodiles
    by Herbert S. Zim
    Pictures by James G. Irving
    William Morrow & Company

This is an older book, but it explains the habits of crocodiles in a clear style. Good for teacher information and for intermediate grade students.

Lyle and the Birthday Party
    by Bernard Waber
    Houghton Mifflin

A family's pet crocodile becomes unhappy because he has never had a birthday party.

A Crocodile's Tail
    by Jose and Ariane Aruego
    Scholastic Books

A Philippine folk story of a boy who rescues a crocodile. Then the boy is rescued by a monkey from the hungry crocodile.

Crocodile Facts:

Most reptiles are silent, but crocodiles make one of the loudest sounds in the animal kingdom.

Crocodiles are the largest living reptiles today.

Only **one** kind is found in the U.S.A. It lives in Florida.

Crocodiles and Alligators are much alike but...

Alligators have broad, rounder heads.

Crocodile's head is narrow and pointed. His eyes stick up farther and he has fewer teeth. His teeth are longer and sharper!

Unbored Bulletin Boards

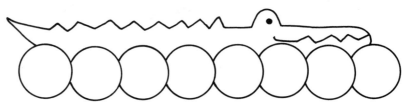

Your good stories keep me afloat!

This is a good way to display work in your room. The alligator is cut from a green strip of butcher paper. Add his eye and teeth with a felt pen. The buoyant bubbles are cut from blue construction paper. The children's papers may be pinned in the circle for display. Their names are put in with black felt pen. The title is cut from black paper.

3-D art stuff — A Bumpy Crocodile

Materials:
- green construction paper
  2 pieces 4" x 18" (10.2 x 20.3 cm)
  1 piece 4" x 4" (10.2 x 10.2 cm)
- white construction paper
  4" x 4" (10.2 x 10.2 cm)

crayons, paste, and scissors

1. Put 2 large green pieces together.

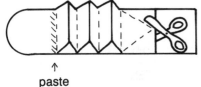

Now fold top paper back and forth:

Now cut tail and paste together.

2. Small green paper for the legs.

Round off end of all 4 legs, then paste in place.

3. White paper for eyes and teeth.

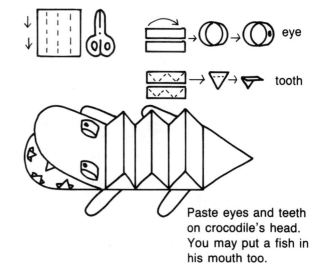

eye

tooth

Paste eyes and teeth on crocodile's head. You may put a fish in his mouth too.

©1979 by EVAN-MOOR CORP.

Draw a crocodile basking in the sun.

Using words from the Word Bank, write about your picture.

Word Bank

long	ferocious	swim	run
scaly	dangerous	eat	snap
rough	sharp	rest	teeth
sturdy	thin	sleep	reptile
hungry	broad	swim	camouflage

A Tale of a Tail

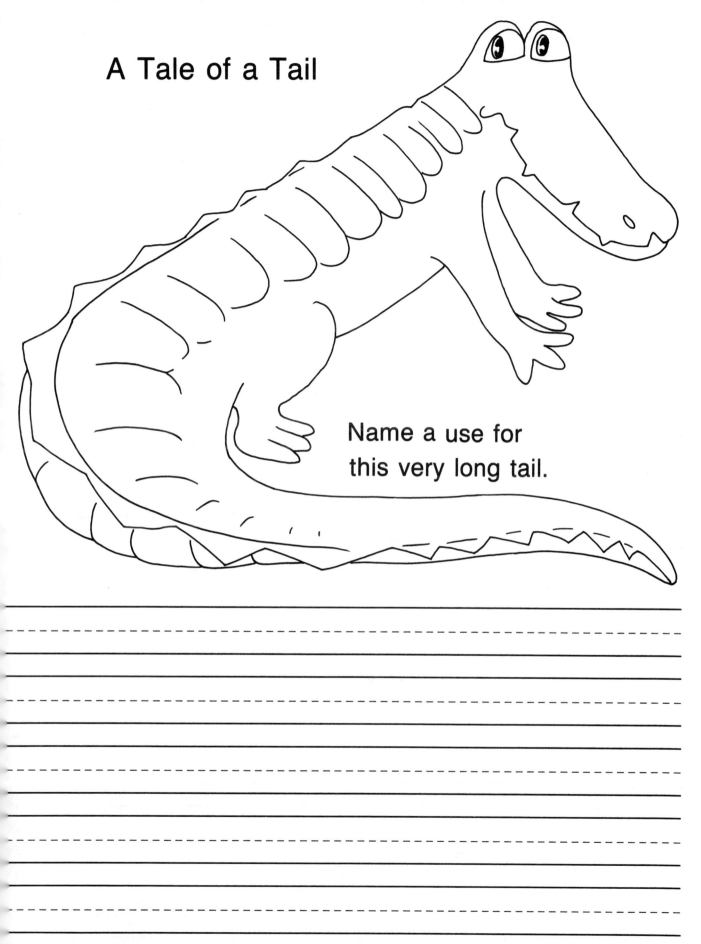

Name a use for this very long tail.

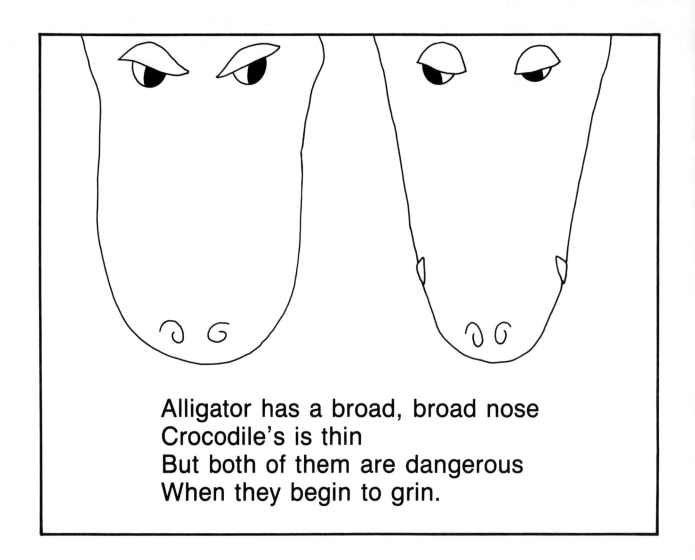

Alligator has a broad, broad nose
Crocodile's is thin
But both of them are dangerous
When they begin to grin.

This poem tells you one way to tell a crocodile from an alligator. See if you can find <u>three more</u> differences. Look in animal books, encyclopedias or dictionaries for help. Try the library.

1. _____

2. _____

3. _____

The Name Game

One child selects a card containing an animal name. The child must then answer all questions asked of him with words beginning with the same letter as the first in the animal's name.

Example:
1. What's your name?
   Connie Crocodile
2. Where do you live?
   Catalina
3. What do you eat?
   carrots and cauliflower
4. What is your job?
   carpenter
5. What do you like to play?
   cards

Mouse Mischief

By now your overhead projector pen may need replacing. You've been using it a lot. If you're using a chalkboard, I hope you're not overcome with dust.

This lesson offers an opportunity to develop several language skills:
- rhyming words (mouse-house, mice-nice)
- plurals (mouse-mice, goose-geese, foot-feet)
- cat and mouse adventure tales

We will give attention in this lesson to differentiating the foreground from the background.

The use of "wiggle lines" will help lend a sense of movement to this mischievous mouse.

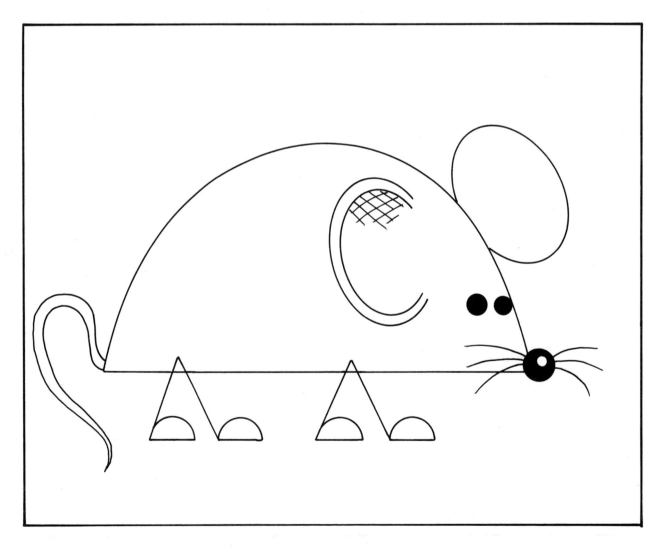

Drawing Steps

Begin with a horizontal line just to the left of center. Follow with a hill above the line.

Add one circle on the right corner for a twitchy nose, two circles above the nose for tiny eyes and six curly whiskers.

Two large circles form mouse's ears. Discuss how a profile view allows you to see the inside of only the one ear. Cross hatching gives you depth.

A twisting tail is a must! "Wiggle lines" give the illusion of movement. His legs move and bend; vary the positions for fun.

Distinguish background from foreground by drawing a horizontal line behind mouse. A hill on that line creates a mouse hole.

Now children may add finishing touches.
- Want to make mouse into MICE?
- Perhaps they'll want to draw appropriate furnishings for a mouse house?

Follow up soon by drawing mouse having a nibble of his favorite cheese.

Meese? Mouses? Mice?
One is just a little mouse
Scurrying around the yard
   or house
But when more than one
   I chance to see
I'm confused!
What should it be?
        J.E. Moore

Unbored Bulletin Boards

Mouse's Favorite Things

This easy-to-make bulletin board is fun to adapt to any area of the curriculum. It is a great way to display children's work or to reinforce science facts, math facts, reading vocabulary, etc. Mouse is cut from black butcher paper and the board itself is backed in mouse's favorite color — "creamy, cheesy yellow."

Literature

Harvest Mouse
   by Oxford Scientific Film
   Pictures by George Bernard
   Putnam

Photographs make it easy to follow the life of a harvest mouse.

The Story of Rodents
   by Dorothy E. Shuttlesworth
   Pictures by Lydia Rosier
   Doubleday and Co., Inc

A beautifully illustrated book containing factual information on many varieties of rodents.

Frederick
   by Leo Lioni
   Pantheon Books

While most mice are storing up food for winter, Frederick stores up colors and words for when the gray days of winter come.

Wild Mouse
   by Irene Brady
   Scribners

The story of the first sixteen days of the life of a group of white-footed mice.

Mouse and Company
   by Lilo Hess
   Charles Scribner's Sons

This book gives many facts about the development and habits of the Deer Mouse. Includes how to keep them for pets. Nice photographs.

Sing Little Mouse
   by Aileen Fisher
   Pictures by Symeon Shimin
   Thomas Y. Crowell Co.

The story of a boy's search for a singing mouse. Written in verse with beautiful illustrations.

 # Mouse Facts

- Mice are more active at night.
- Meadow Mouse eats its own weight in seeds every 24 hours.
- Field mice use their tails for climbing and balancing.
- White-Footed Mouse hums in its throat and thumps its front paw.
- Meadow Jumping Mouse can jump twelve feet.
- Hazel Mouse hibernates in the winter.
- Mice are rodents!
- Baby mice are born blind and hairless.

Language Art Starts

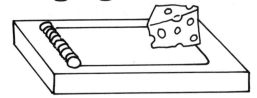

Let's Talk Traps!

What would you want to trap?
Would you invent your own special trap? Describe it!
Where would you leave the trap?
How would you bait the trap?
Have you ever been "trapped"?

3-D art stuff — Mouse Puppet

Read the "City and Country Mouse" tale and then make two of these puppets to depict the story!

Materials:
- one 9" x 12" (22.9 x 30.5 cm) gray construction paper
- two 4" x 4" (10.1 x 10.1 cm) gray
- one 6" x 3" (15.2 x 7.6 cm) black
- one 6" x 3" (15.2 x 7.6 cm) red or blue
- two 3" x 3" (7.6 x 7.6 cm) pink
- scissors, paste and crayons

1. Fold 9" x 12" gray paper into thirds. Overlap and paste. Round off top for head.

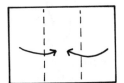

2. Paste top of head closed. Cut ears from the 4" x 4" gray and 3" x 3" pink. Paste pink on gray. Cut a slit in the bottom of ears. Overlap slightly and paste for a 3-D look.

3. Cut eyes, nose, whiskers and tail from black. Paste them on and then draw the mouth and red cheeks.

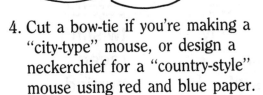

4. Cut a bow-tie if you're making a "city-type" mouse, or design a neckerchief for a "country-style" mouse using red and blue paper.

Draw a mouse nibbling at its favorite cheese.

Using words from the Word Bank, write about your picture.

Word Bank

gray	cautious	escape	mouse
brown	careful	nibble	mice
twitchy	twisting	cheese	rodent
tiny	raced	hole	whiskers
swift	hurried	trap	tail

Hungry cat meets plump mouse...
What do you think will happen next?

_____

Circle your answer:
   Did the hungry cat trap its dinner?   yes   no
   Did the plump mouse escape?   yes   no
   Did the cat get away?   yes   no
   Is your story —

      funny   exciting   unusual

One is just a little mouse
Scurrying around the yard or house
But when more than one
   I chance to see
   I'm confused
      What should it be?

Meese?
Mouses?
Mice?

one | more than one

1. mouse

2. goose

3. deer

4. horse

5. sheep

6. fox

7. octopus

8. ox

Pick-A-Pair

This activity provides practice in matching singular and plural forms. You will need tag strips containing three words (a singular form, its plural and one "similar" word) numbered 1, 2, 3.

Example:

1. mice	2. meese	3. mouse

Children read the three words silently and raise fingers to indicate the word that does not belong (1 finger, 2 fingers or 3 fingers). The teacher then covers the incorrect word and children recite the correct answers in sentences.

Mouse means one.
Mice means more than one.

1. goose 2. gooses 3. geese	1. ox 2. oxen 3. oxes	1. mans 2. man 3. men	1. foxen 2. fox 3. foxes
1. children 2. childs 3. child	1. half 2. haves 3. halves	1. feet 2. feets 3. foot	1. sheeps 2. sheep 3. sheep

Walrus Whimsy

Walrus is easy to draw and is a unit that is enthusiastically enjoyed by children of all ages. He is a lively topic for comic art and fictitious tales of fun and daring.

The science facts about walrus are fascinating. The idea that a body so encumbered could move with such speed is a wonder.

Follow the regular procedures for development of the drawing as well as vocabulary and fact enrichment.

Be sure Walrus is drawn large enough to fill the page!

Drawing Steps

POST TEST

Begin with a tall hill left of center. Allow space at the lower edge for his bottom flipper. Draw a line across the top to form the head.

Add a snout on the top line. Draw the nose and tusks. The bottom flipper reaches to the side.

The bulging black eyes are looking straight ahead at us! Draw bristles for the moustache.

The walrus has a double-double chin. Draw flippers on each side.

Walrus rests happily on an ice floe.

Try outlining the wave formations with the cool colors: blue, green and purple. Outline walrus in black and color his body brown.

Ask children to draw walrus swimming in the water searching for clams.

A bristly moustache across his face
Great white tusks firmly in place
Weighing a ton at the very least
Walrus is a marvelous beast.

J.E. Moore

Language Art Starts

Walrus is a good subject for the creation of a cinquain.
Cinquain is a five-lined poem and each line follows a form:

1. One word (title)

2. Two words (describing the title)

3. Three words (describing an action)

4. Four words (expressing a feeling)

5. One word (referring back to the title)

Practice as a group — then explore!

Literature

Seals, Sea Lions & Walruses
   A Zoobook
   by John Bonnett Wexo
   Wildlife Education, Ltd.

Excellent photographs and illustrations describe the physical make-up and habits of the pinnipeds.

The Walrus and the Carpenter
   by Lewis Carroll
   Pictures by Gerald Rose
   E.P. Dutton & Co.

Contains all the verses of *The Walrus and the Carpenter* plus many other poems by Lewis Carroll.

The Walrus, Giant of the Arctic
   by Kay McDearmon
   Dodd, Mead and Co.

This book follows the travels of a mother walrus and her year-old calf. Illustrated with black and white photographs.

Walpole
   by Syd Hoff
   Harper & Row

An easy reader about the biggest and strongest walrus of all and how he becomes leader of the herd.

Seals of the World
   by Gavin Maxwell
   Constable and Company

An excellent source for teacher research.

©1979 by EVAN-MOOR CORP.

Walrus Facts:

Unbored Bulletin Boards
Sharing Center

He has no outside ears.

Male tusks may be 3 feet long!

He may weigh 3000 lbs.

He lives in the Arctic among icebergs.

He uses tusks to dig clams.

He has no fur but 2" thick skin, and 6" of blubber.

He has about 400 bristles in his moustache which serves as a sensory function and helps shovel food in.

"The time has come," the Walrus said, "to speak of many things!

Arrange a table below this board so the children can display articles or books they've brought to school to share. You can jot down a sentence for each item shared on a tag sentence strip and lay it on the table. Want to develop a discussion on the use of quotation marks?

Tara said, "This lava rock is light."

"I wanted you to see this book," said Gus.

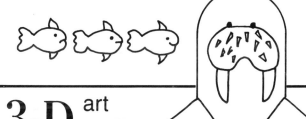

3-D art stuff

Walrus Puppet

Materials:
- paper: 9" x 12" (22.9 x 30.5 cm) light brown
  8" x 3" (20.3 x 7.6 cm) dark brown
  2" x 3" (5 x 7.6 cm) white
- 8 flat toothpicks (or 4 broken in half)
- paste, glue, scissors, crayons

1. Fold light brown paper into thirds. Overlap and paste. Round off top for head. Paste top of head closed.

2. Fold 8" x 3" dark brown paper and cut on fold lines.

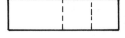

moustache   flippers.

3. Fold 2" x 3" white paper and cut on the fold line. Round off one end for tusks.

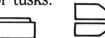

4. Assemble puppet.

Paste on moustache.   Draw eyes and nostrils in black crayon.   Paste on tusks and flippers.

5. Glue toothpicks onto moustache for bristles.

Allow glue to <u>dry</u> before using.

Now...enjoy him!

Draw a walrus swimming in the water searching for clams.

Using words from the Word Bank, write about your picture.

Word Bank

wiggly			flippers
lumpy	double chin	arctic	moustache
bulging	lumber	iceberg	tusks
bristly	crunch	blubber	ocean
thick	float	whiskers	dive

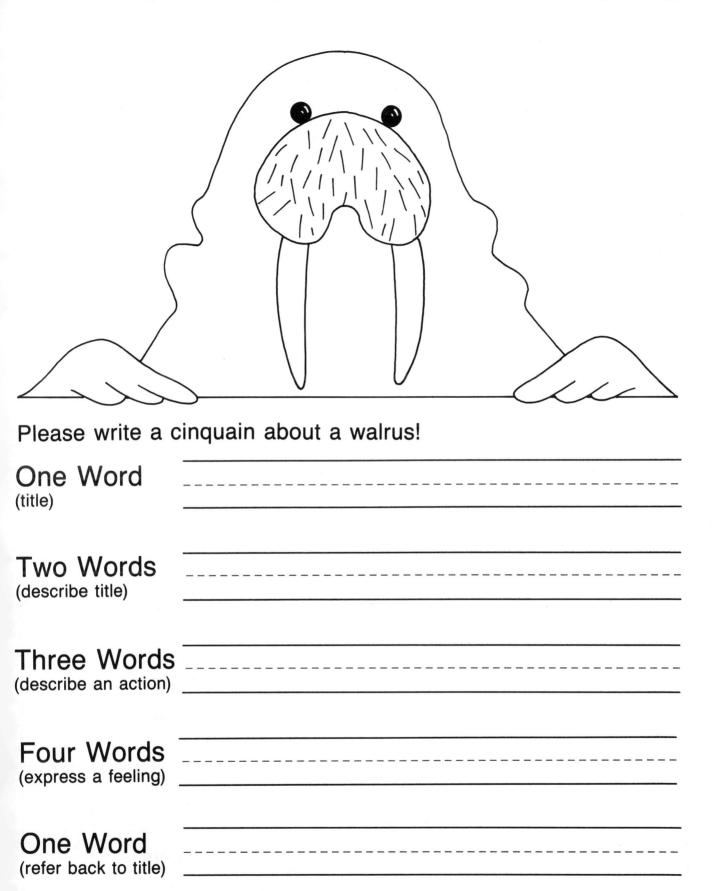

Please write a cinquain about a walrus!

One Word
(title)

Two Words
(describe title)

Three Words
(describe an action)

Four Words
(express a feeling)

One Word
(refer back to title)

Can you read your cinquain to a friend?

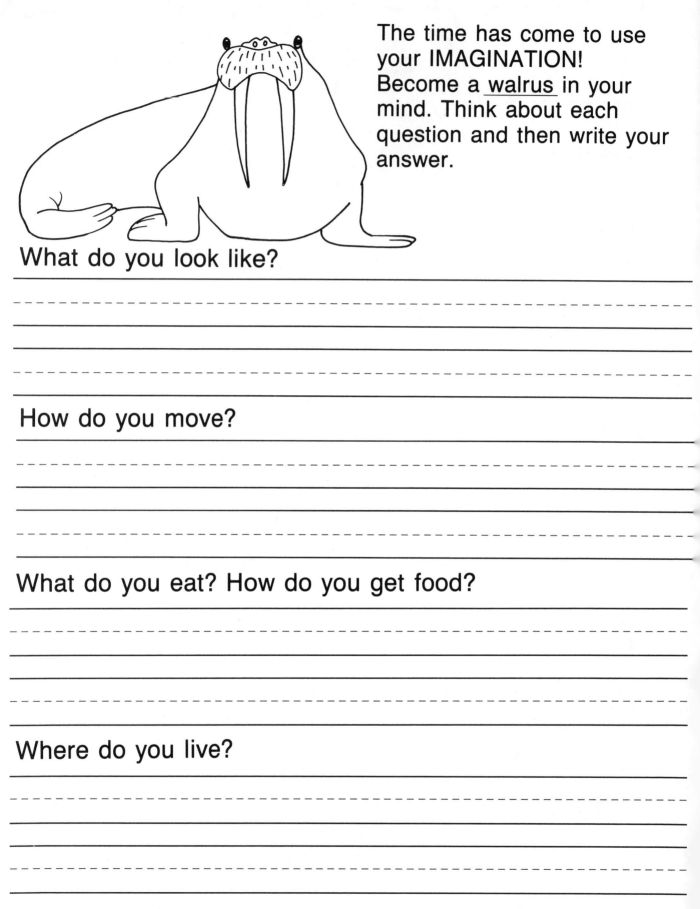

The time has come to use your IMAGINATION! Become a <u>walrus</u> in your mind. Think about each question and then write your answer.

What do you look like?
_____
_____
_____
_____

How do you move?
_____
_____
_____
_____

What do you eat? How do you get food?
_____
_____
_____

Where do you live?
_____
_____
_____
_____

Now draw yourself on the back of this paper. Thank you.

Is it true?

Thumbs up for true facts.

Thumbs down if facts are false.

1. A walrus has no outside ears. TRUE
2. Walrus tusks can be 3 feet long. TRUE
3. A walrus is covered with fur. FALSE
4. A walrus can move as fast as a man. TRUE
5. A baby walrus rides in a pouch on his mother. FALSE
6. A walrus has skin 2 inches thick. TRUE
7. Walruses eat clams. TRUE
8. Female walruses do not have whiskers. FALSE
9. A walrus can breathe under water. FALSE
10. A walrus can swim 15 miles per hour. TRUE

Bear Fare

The bear can be adapted by children to draw and create original stories and illustrations about the many popular bear characters in literature.

Science and vocabulary development can benefit from a discussion of hibernation and other bear habits.

Texture is explored here with the lines used to simulate fur.

Children need to be reminded that sketching means using the pencil with a very light and carefree touch.

Bears need to be LARGE to be effective.

Drawing Steps

Begin with a centered horizontal line. Add a hill above the line. A circle sits over the hill.

Bear's ears are double circles. Another hill forms the tummy. The legs and feet extend to the side.

Draw a circle for the facial mask. The arms reach out and up.

Two sparkling eyes appear. Add a nose that slips easily into a smile.

Create a "furry illusion" by using short slash lines around the outside.

What can individualize your drawing?
- Is bear picking a bouquet of wildflowers?
- Is bear's brother hiding behind a tree?

Follow up soon by asking children to draw mother bear and her two cubs.

Bears Bears Everywhere Bears

Grizzly bears are daring and bold
Polar bears live in Arctic cold
Black bears busily search for grubs
Sun bears play with their baby cubs
Wherever you go the whole world round
The mighty bear family can be found

J.E. Moore

Language Art Starts

- Let them recreate the Three Bears story.
- Memorize the poem.
- Have each "personal bear" describe his very favorite lunch menu.
- Answer the question: "What's the bruin doing next winter?"

Use the bear puppets to encourage oral creativity.

Literature

Bears and *Polar Bears*
  by John Bonnet Wexo
  Wildlife Education, Ltd.

Both of these Zoobooks contain excellent photographs and illustrations and clearly written texts describing bears and their habits.

Black Bear Baby
  by Berniece Freschet
  Pictures by Jim Arnosky
  Putnam

An easy-to-read description of the first months of a baby black bear's life.

A Kiss for Little Bear
  by Else H. Minarik
  Pictures by Maurice Sendak
  Harper and Row

A delightful story about Little Bear who sends his grandmother a gift and receives a kiss in return.

Wild Babies, A Canyon Sketchbook
  by Irene Brady
  Houghton

Lovely sketches by the author illustrate information about the development and growth of baby bears and other common wildlife.

Blueberries for Sal
  by Robert McCloskey
  Viking

This charming old story follows the adventures of a little girl and a baby bear while out gathering blueberries with their mothers.

Baby Bears and How They Grow
  by Jane H. Buxton
  National Geographic

An excellent book to introduce bears, their life cycle, and habits to young students.

©1979 by EVAN-MOOR CORP.

Bear Facts

Bears have poor eyesight but a keen sense of smell and hearing.

The bear walks awkwardly; he lifts both right feet, then left.

They can run very fast for short distances.

Bears have 5 toes on each foot with a <u>sharp</u> claw on each toe.

Bears sleep during winter. They store fat under their skin for warmth and nourishment while asleep.

Bears eat many things: meat, fish, berries, grubs, ants, honey.

Unbored Bulletin Boards

I spend all winter with a book!

Try this in your library center. You may display real book jackets and change weekly. It's also a good motivational technique to post children's stories in bear's paws. Bear may also be used with a different title to emphasize current math or language concepts. He's versatile and easy to make.

3-D art stuff

Here's a Cuddly Bear Puppet

Materials:
- paper: brown 12" x 18" (30.5 x 45.7 cm)

  Bear's body

  pink 3" x 18" (7.6 x 45.7 cm) Cut 3" squares
  black 1½" x 12" (3.8 x 30.5 cm)
  Cut off 3–1½" (3.8 cm) squares
- two brass fasteners
- paste, scissors, crayons

1. Take two 4" brown squares and the six pink 3" squares and round off corners to make circles.

2. Cut two of the pink circles in half for bottoms of hands and feet.

3. Round off top corners of brown 5" x 4" pieces. Paste on pink ½ circles.

4. Fold the bears body in half. Then fold the top piece back. Cut the legs up the center as far as the second fold. Paste on brown ears.

5. Paste pink circles:

6. Take the three 1½" black squares and round off corners. Paste these on for bear's nose and eyes. Draw a smile with a crayon.

7. Lay the arms in place; insert fasteners. Flip bear over and put black (1½" x 7½") strip over fasteners.

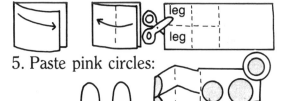

©1979 by EVAN-MOOR CORP.   Animal Units

Draw mother bear and her two cubs.

Using words from the Word Bank, write about your picture.

Word Bank

large	dig	zoo	bruin
daring	hunt	grubs	cub
fast	sleep	berries	claw
awkward	climb	honey	woods
hairy	smell	cave	dangerous

Please help Mr. Bear plan today's menu by writing in your delicious ideas.

Breakfast for the Three Bears:
1. _____
2. _____
3. _____

Lunch with Little Bear:
1. _____
2. _____
3. _____

Tea for Pooh Bear:
1. _____
2. _____
3. _____

Dinner with Ms. Polar Bear:
1. _____
2. _____
3. _____

Oh, oh! Here comes Grizzly Bear!
What can Mr. Bear feed him?

What is Bear dreaming about during his long winter nap?

Rhyme Around the Circle

The children sit in a circle. The teacher names an animal. Each child in the circle gives a word rhyming with the animal named.

cat	hat, bat, fat, sat, etc.
snail	pail, trail, mail, etc.
bear	pear, care, stair, etc.
cow	how, now, plow, etc.
frog	log, dog, smog, hog, etc.
mice	nice, price, rice, spice, etc.
pig	big, dig, fig, wig, etc.
duck	luck, pluck, cluck, etc.
hen	pen, ten, wren, den, etc.
snake	lake, rake, bake, flake, etc.

This is an excellent time to read **Bears** by Ruth Krauss. It is all written in rhyme.

Elephant Illustrations

Get your overhead pen ready. This one is always a HUGE success! Elephant prose and poetry sources abound and children are very responsive to creating stories and illustrations on this subject. A unit on elephants can utilize reading, language arts, music, art, math and science topics for a unified study enjoyed by all.

We continue to emphasize:
- sketching lightly with pencil
- draw large enough to fill paper
- utilize descriptive vocabulary

This lesson will also concentrate on important directional instructions and vocabulary: above, below, over, under, etc.

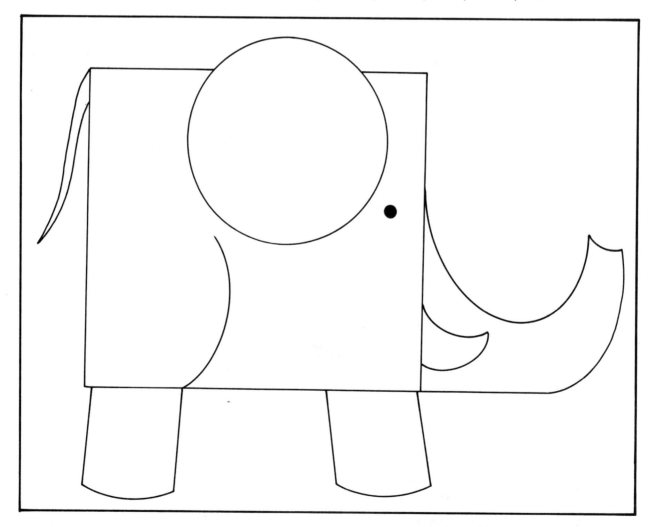

Drawing Steps

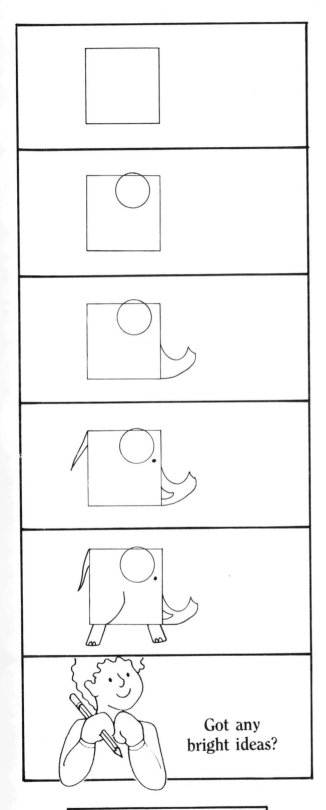

Draw a large square. It should be offset toward the left side of the paper.

A circle forms elephant's ear. Only one ear is visible because of the profile view. The circle is in the upper right-hand corner.

My trunk swings upward looking for peanuts. Where is yours? Variety is allowed!

A small eye appears below the large floppy ear and a tusk beside the trunk. A tail ready to swish brings up the rear.

Two short stubby legs are drawn on the bottom of the rectangle. The rear leg begins high on the left. Toenails add interest.

Where does your elephant live?
- in a jungle
- in the zoo
- with a circus

Follow up tomorrow by asking students to draw an elephant reaching for a delicious peanut.

Language Art Starts

Gray
Wrinkled
Saggy
Skin
that's the cover
an elephant's in.

J.E. Moore

Way down south
where bananas grow
A grasshopper stepped
on an elephant's toe.
The elephant cried
with tears in his eyes
Pick on somebody
your own size.

Anonymous

"I have small ears. I have only one ''finger-like'' tip, but I can still pick up objects."

"I have very large ears and tusks. I have two ''finger-like'' tips on the end of my trunk to help me pick up things."

Let's write about the 2 types

Here is a TUSK-FULL of big words to use in writing: gargantuan, enormous, gigantic, immense, huge.

Literature

Elephant Facts
 by Bob Barner
 E.P. Dutton

Humorous cartoon sketches and a simple text teaching interesting facts about elephants.

Elephants
 by Elsa Posell
 Childrens Press

A good introduction to elephants with interesting illustrations.

Little Wild Elephants
 by Anna Michel
 Pictures by Peter &
 Virginia Parnall
 Pantheon

The first four years of an elephant's life is told with easy text.

Elephants
 by Joe Wormer
 E.P. Dutton

This book compares Asian and African elephants using large photographs.

Elephant Fact Test

Here is a quiz to stimulate interest in the beginning of the unit. Children indicate their choice of answer by thumbs up for true, thumbs down for false.

What do YOU know about elephants?
TRUE or FALSE

F 1. Most elephants come from Africa.
F 2. A man can outrun an elephant.
F 3. Elephants are the largest of all animals.
F 4. An elephant's trunk is its best weapon.
T 5. Tusks grow back if they are broken.
F 6. An elephant's trunk has only one use.
T 7. An elephant's skin is an inch or more thick.
T 8. Elephants can't run or gallop.
F 9. Elephants are afraid of mice.
T 10. An elephant can smell water three miles away.

Unbored Bulletin Boards

YOU are a huge success.

Make this fanciful elephant head out of the classified ad section of your newspaper. Cut the paper to fit the dimensions of your board. The backing should be a bright color to provide contrast. Make the "bubble" on white paper. The cheeks are red circles. The trunk is a strip of paper pleated like an accordion. Now pin up children's work on his ears.

3-D art stuff   Stand-Up Elephant Parade

1.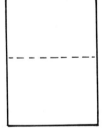

Pick a color you like of 9" x 12" (22.9 x 30.5 cm) construction paper. Fold it in half.

2.

Make 1 cut for the trunk. Cut out a half circle on the bottom and round off the top right-hand corner.

3.

Use the half circle scraps for the two ears. Use the rounded off corners for the tail. Paste on.

4.

Draw eyes and toenails with crayons or felt pens. Fold up feet each a ½". Stand him on his own four feet!

Make several in different colors and sizes of paper. Crayons can add bright circus blankets.

Draw an elephant reaching for a delicious peanut.

Using words from the Word Bank, write about your picture.

Word Bank

enormous	gray	swish	tusks
rough	stubby	swing	skin
huge	heavy	work	tail
agile	thick	lift	trunk
fast	floppy	trumpet	jungle

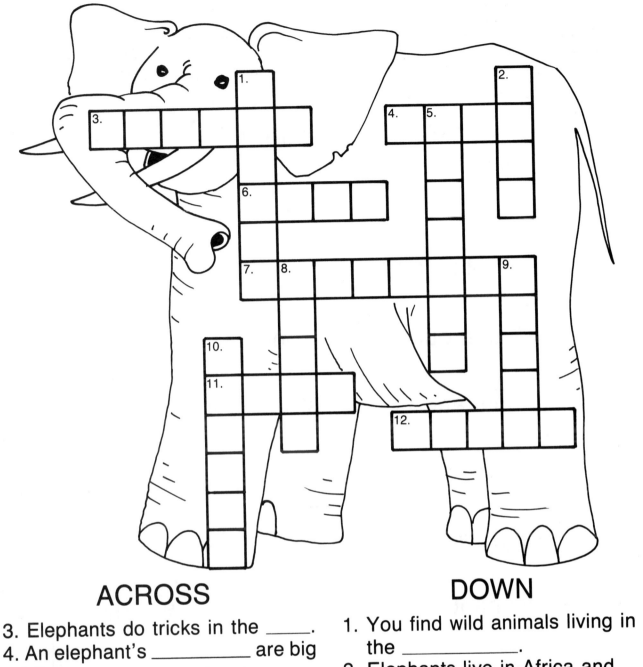

ACROSS

3. Elephants do tricks in the ____.
4. An elephant's _____ are big and floppy.
6. The color of an elephant is____.
7. A large animal with a trunk.
11. An elephant has four _____.
12. Two large curved teeth sticking out of an elephant's mouth.

DOWN

1. You find wild animals living in the _____.
2. Elephants live in Africa and ___.
5. Cows, ants and bears are all _.
8. Mice are small. Elephants are_.
9. An elephant can pick up things with its _____.
10. Elephants eat _____.

Word Bank

tusks	elephant	legs	gray
Asia	jungle	large	animals
ears	plants	circus	trunk

©1979 by EVAN-MOOR CORP. Animal Units

What do you know about elephants?

Write: T for true
F for false

1. Most elephants come from Africa.

2. A man can outrun an elephant.

3. Elephants are the largest of all animals.

4. An elephant's trunk is its best weapon.

5. Tusks grow back if they are broken.

6. An elephant's trunk has only one use.

7. An elephant's skin is an inch or more thick.

8. Elephants can't run or gallop.

9. Elephants are afraid of mice.

10. An elephant can smell water three miles away.

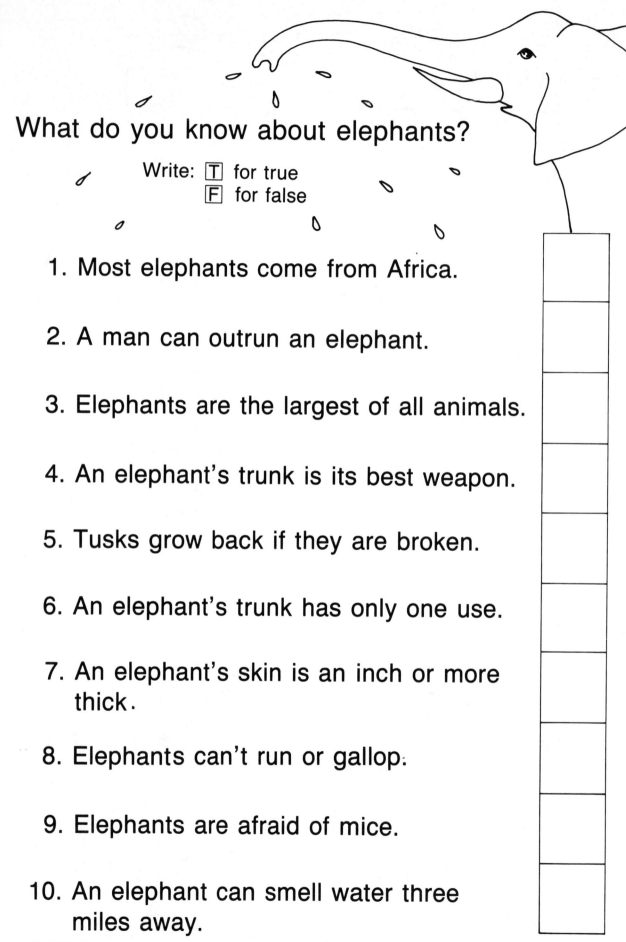

Animal Units

I'm a Word Hunter!

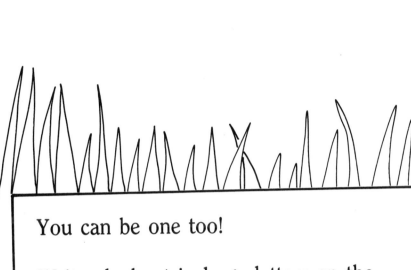

You can be one too!

Write elephant in large letters on the chalkboard. Children see how many words they can make using the letters in elephant. How about a 5-minute time limit?

Examples:

ant	eel	hat
tan	plant	leap

How about peanuts for the winners!

Elephant

Giraffe Acts

You won't have any trouble convincing children to draw large and fill the paper on this one!

The "giraffe trio" offers you a choice. You may begin by drawing only one giraffe or they may be drawn in a series to simulate the motions of bending down. Children enjoy creating flip books to show this action speeded up. Giraffe is beautiful whether he is moving or standing still.

Design possibilities may be discussed when considering the way giraffe's spots seem to "just fit together."

The lanky giraffe also offers humorous veins to explore in drawing and story-telling.

Drawing Steps

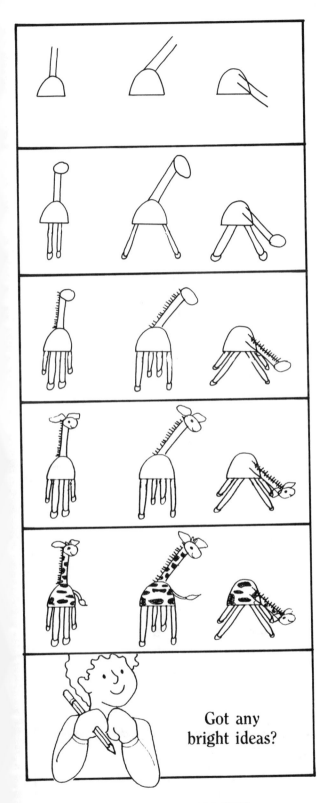

Begin with the half-circles spaced across the paper. The neck columns reach almost to the top edge.

The oval heads may turn other directions. Giraffe's legs are shown in various positions. Bend and move then in your own way.

The rear legs lend a comic appeal. Add fringe along the neck.

Draw the ears. You may choose to show both ears in your pose. Add a nose and mouth.

Draw two horns for each giraffe. Does your giraffe have a swishing tail? Giraffe spots are brown and fit together.

Now where can you go with your drawing?

Where can a guy get a juicy tree and drink of river water?

I'm stunning outlined in black!

Follow up by challenging the class to draw giraffe on a piece of construction paper cut to 6" x 18".

Got any bright ideas?

POST TEST

Language Art Starts

Hey, Giraffe!
What's it like up there?
What do you see
With your head in the air?

Giraffe doesn't answer.
He stands under the trees,
Looks around quietly,
And nibbles on leaves.
  J.E. Moore

Category Game

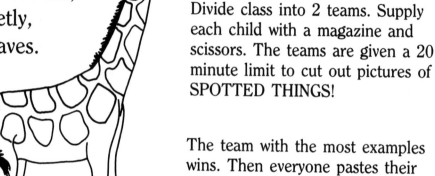

Divide class into 2 teams. Supply each child with a magazine and scissors. The teams are given a 20 minute limit to cut out pictures of SPOTTED THINGS!

The team with the most examples wins. Then everyone pastes their findings on a large chart and labels their discoveries. (Freckled people excel at this task.)

Literature

Giraffe Lives in Africa
   by Gladys Conklin
   Pictures by M. Kalmenoff
   Holiday House

The experiences of a baby giraffe growing up written in a story form.

Giraffes
   by John Bonnet Wexo
   Wildlife

This Zoobook contains the answers to all sorts of questions about giraffes: how they are built, what they eat, their history, etc. Excellent photographs and illustrations.

Giraffe, the Silent Giant
   by Miriam Schlein
   Pictures by Betty Fraser
   Four Winds Press

For teacher information or intermediate students. Contains everything you ever wanted to know about giraffes.

Daisy Rothschild: The Giraffe that Lives With Me
   by Betty Leslie-Melville
   Doubleday

Full-color photographs help children learn about a baby giraffe raised by humans.

Giraffe Facts:

Unbored Bulletin Boards

Some can reach the height of eighteen feet!

Their bodies are short. The height comes from long necks and legs.

Giraffe's neck has only seven vertebrae.

Their front legs are longer than their back.

They are native to Africa, south of the Sahara, and live in open bush country.

Giraffes browse upon trees.

A giraffe's full gallop = 30 m.p.h.

They do make sounds, but are seldom heard.

They are the tallest mammals.

Each of my spots is unique! What makes you unique?

Each child is given "spot space" to write about their capabilities. The spots cover the whole board. Giraffe's head and legs are taped to the walls above and below to make giraffe very tall.

3-D art stuff

Materials:
- paper —
  one yellow 12" x 18" (30.5 x 45.7 cm) (backing)
  one yellow 10" x 4" (25.4 x 10.2 cm) (neck)
  one brown 6" x 9" (15.2 x 22.9 cm) (tree and spots)
  green tissue in 1" squares (2.5 cm)
  one black 1" x 12" (2.5 x 30.5 cm)
- paste, scissors, crayons

1. Cut tree trunk off edge of brown paper. Paste it on left edge of large yellow.

2. Crush tissue paper squares over eraser end of pencil and paste on for leaves.

Giant Giraffe Stretches His Neck

3. Draw giraffe's head and neck on narrow yellow strip. Outline in black and draw in giraffe's face.

4. Cut black strip.

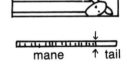
mane   ↑ tail

Paste mane to neck and horns to the head.

5. Paste the head to the large paper. Sketch the rest of giraffe's body directly on the yellow paper and outline with black crayon.

6. Cut free-form brown spots and paste on body and neck of giraffe.

7. Draw in the background with crayons.

1. Draw a very tall giraffe.

2. Write about this giraffe.

Word Bank

tall	bony	swish	plains
elegant	quiet	browse	leaves
spotted	move	bend	tongue
shy	reach	legs	horns

Find the missing animal!
Fill in the animal names.
Read down the darker boxes to find the missing animal.

Draw the missing animal on the back of this paper!

Word Bank

1.
2.
3. (turtle)
4.
5. (frog)
6. (elephant)
7.
8. (bear)
9.

giraffe   snail   frog
walrus   crocodile   elephant
turtle   bear   mouse

**Giraffe is a unique animal.
What is unique about YOU?**

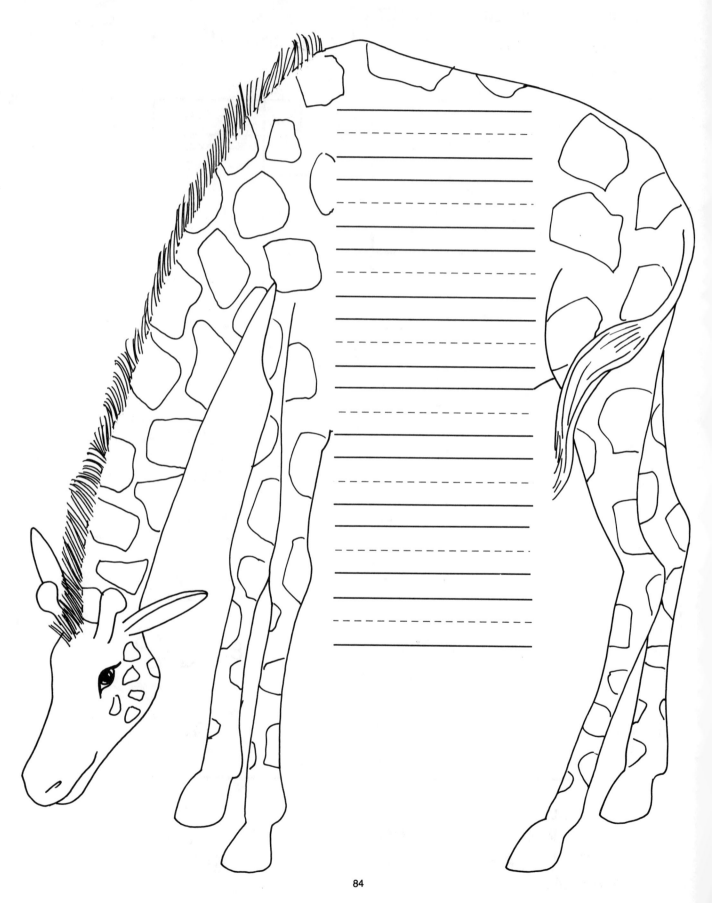

Bean-Bag Toss — Opposites

Make a playing board on tag or oil cloth with permanent pens. Draw circles all over the board. Put one word in each circle, being sure to include an opposite for each word used. You will also need two bean-bags.

A child stands behind a masking tape line and tosses one bag. He reads the word the bag lands on and tries to make the second bean-bag land on its opposite.

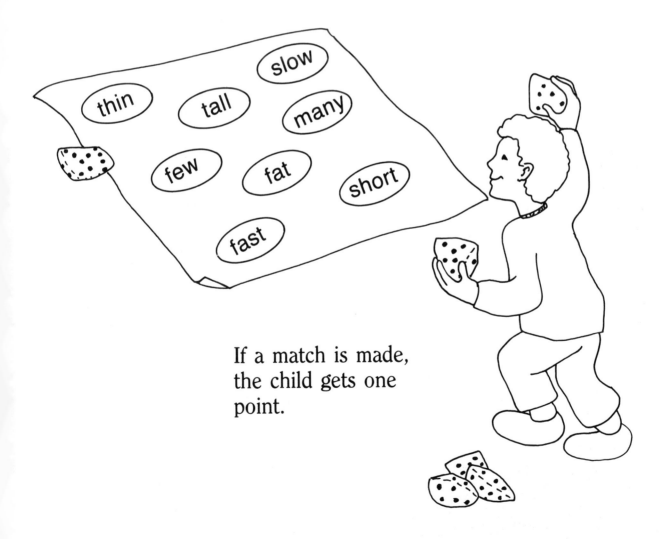

If a match is made, the child gets one point.